Helping Smokers QUIT

The Science Behind Tobacco Harm Reduction

Presented by the

AMERICAN COUNCIL ON SCIENCE AND HEALTH

Based on a publication by **Brad Rodu**

December 2011

ACSH accepts unrestricted grants on the condition that it is solely responsible for the conduct of its research and the dissemination of its work to the public. The organization does not perform proprietary research, nor does it accept support from individual corporations for specific research projects. All contributions to ACSH—a publicly funded organization under Section 501(c)(3) of the Internal Revenue Code—are tax deductible.

Individual copies of this report are available at a cost of $6.50. Reduced prices for 10 or more copies are available upon request.

Published by the American Council on Science and Health, Inc.—December 2011.

Contents

Introduction

In 2006, the American Council on Science and Health (ACSH) became the first organization in the United States to formally endorse tobacco harm reduction (THR). ACSH based its position on a comprehensive review of the existing scientific and medical literature, which showed that smokeless tobacco use:

(a) is at least 98 percent safer than smoking, even though most Americans are misinformed about the differences in risk

(b) among Swedish men, is a major factor in extremely low smoking rates

(c) is not a gateway to smoking cigarettes

Over the past 5 years the scientific literature supporting THR has grown considerably, and ACSH decided to sponsor another comprehensive review. This publication summarizes the major findings of that review.

Smokeless Tobacco, Small Risks

Unlike cigarettes, smokeless tobacco products are smoke-free. Users place small pouches between the upper lip and gum, thereby avoiding the thousands of toxic agents formed when tobacco is burned.

In the past five years, numerous epidemiological studies have confirmed that the use of smokeless tobacco is associated with minimal risks for cancer and heart disease and has no risk for gastrointestinal disorders or chronic inflammatory disease.

Cancer

A comprehensive 2009 review found that smokeless tobacco use was not associated with cancer after adjustment for the effects of smoking. The only exception was a small risk for prostate cancer, although the biologic basis for this risk is not clear, and more information is needed before any definite conclusion can be drawn.

Cardiovascular Diseases

At least 10 epidemiological studies, as well as two meta-analyses, have evaluated the risks for cardiovascular diseases among smokeless tobacco users. The majority of these studies found no significant risks. In one study that found slightly increased risks for stroke and heart attack, the results had been inadequately adjusted for smoking.

In 2010, the American Heart Association released a policy statement based on a review of the scientific literature, which confirmed that smokeless tobacco use confers low risks for cardiovascular diseases.

Gastrointestinal Disorders

Smokeless tobacco use inevitably results in swallowing saliva containing tobacco extract, but in 2010 a Swedish study found no increased risk of GI disorders in smokeless tobacco users. In fact, smokeless users reported GI symptoms with the same frequency as non-users. In contrast, smokers and dual users (those who both smoked and used smokeless tobacco) were more likely to experience dyspepsia.

Parkinson's Disease and Multiple Sclerosis

Smokeless tobacco is not at associated with the neurodegenerative diseases Parkinson's and multiple sclerosis (MS). Intriguingly, smokeless tobacco users actually have lower rates of these diseases than do non-users of tobacco, but a biological mechanism for any possibly protective effect has yet to be explained.

Chronic Inflammatory Disease

Smokeless tobacco is not associated with chronic inflammatory disease. A 2010 study that examined the incidence of these diseases among nearly 280,000 Swedish construction workers found that, compared with those who had never used tobacco, smokers were at significantly greater risk for rheumatoid arthritis, Crohn's disease, and multiple sclerosis, while snus users were not at greater risk. The researchers also conjectured that the inhaled non-nicotinic components of cigarette smoke are more significant than nicotine itself in the etiology of these diseases.

Pregnancy Complications

One of the most challenging questions regarding THR is whether it is applicable to pregnant women who smoke. Our publication, ***Cigarettes: What the Warning Label Doesn't Tell You,*** discusses the spectrum of smoking-related risks to the developing baby and mother thoroughly (pages 118-121). Officially, the Surgeon General has established that smoking during pregnancy is associated with increased risks for premature delivery, low-birth weight, and stillbirth, as well as a number of placental problems that can place both mother and fetus at risk. Can a pregnant smoker who switches to smokeless tobacco benefit the health of her developing baby?

The answer is that, while risks to a developing baby are not as great when a mother uses smokeless tobacco instead of cigarettes, risks do remain. Women who use smokeless tobacco are at risk for slightly smaller babies (an average of six ounces less at birth), as well as modestly elevated risks for premature delivery, stillbirth, and —possibly — preeclampsia. Although any form of nicotine should be avoided during pregnancy, the highest risks for the developing baby are associated with smoking.

Misperceptions About the Minimal Health Risks of Smokeless Tobacco

There is scientific consensus that smokeless tobacco use is vastly safer than smoking, but this is virtually unknown among the general public, and even among health professionals. Indeed, recent studies have revealed the extent of these misperceptions, as well as their potential impact on the implementation of THR.

A 2010 study documented the widespread misperception of smokeless tobacco risks among highly educated university faculty at the University of Louisville. The researchers found that over half of all faculty incorrectly believed that smokeless tobacco poses general health risks that are equal to or greater than smoking. And there was a more common misperception about the risk of cancer associated with smokeless tobacco: Health science faculty, too, were also misinformed about this risk. Overall, the survey demonstrates that most health professionals have a poor understanding of the benefits of smokeless tobacco and are not aware that using it is vastly safer than smoking cigarettes.

Misperception of the risks of smokeless tobacco use clearly stem from a campaign of misinformation by anti-tobacco activists who are more concerned with promoting nicotine abstinence than with a realistic approach to public health. They conflate the risks of smokeless tobacco with those associated with cigarettes, a message relayed via both direct statements and insinuation. Such misperceptions are deeply ingrained: A 2010 study found that even in Sweden, where THR has had a significant impact on smoking, the majority of smokers have an exaggerated perception of the harmfulness of pharmaceutical nicotine and snus.

Undoubtedly, given the actual benefits of switching from cigarettes to smokeless tobacco, greater effort needs to be made to promote accurate perception of the considerably lower health risks of smokeless tobacco.

Smokeless Tobacco is an Effective Substitute for Cigarettes

Evidence from Sweden and Norway makes a strong case for smokeless tobacco as a means to reduce smoking rates.

A 2006 review of smokeless tobacco use in Sweden found that snus was the most common smoking cessation aid among men. The review also found that 58% of Swedish male smokers had used snus as a quit-smoking aid, and two-thirds of them were successful, which was significantly higher than any other aid such as nicotine gum or a nicotine patch.

Allaying fears that smokeless tobacco use was a gateway to smoking, the 2006 review reported that the odds of smoking were significantly lower for men who had used snus than for those who had not. Another study in 2008 found that the use of snus was the strongest indicator of former (versus current) smoking among Swedish men, and it concluded that ""Swedes appear to be using snus as a form of nicotine replacement therapy despite a lack of clinical trials data to support its use as a smoking cessation aid."

Recent reports from Norway have reached similar conclusions. The Norwegian Institute for Alcohol and Drug Research (SIRUS) reported that the prevalence of smoking in young Norwegian men had declined from 50 percent in 1985 to 30 percent in 2007, while the use of snus increased from 10 percent to 30 percent.

SIRUS reported on the popularity of snus as a smoking cessation aid among Norwegian men; it was used by 23 percent of the men who quit smoking in 2007, compared with use rates of 2 to 9 percent for nicotine gum, nicotine patch, the quit-smoking drug Zyban, or quit lines. In 2010 a SIRUS study found that Norwegian men prefer snus over all other methods to quit smoking.

Among Norwegian men who had tried snus as a smoking cessation aid, 74 percent reported to have "quit smoking altogether," or to have experienced a "dramatic reduction in smoking intensity" — a success rate significantly higher than the 40 to 50 percent success rate of men who tried to quit using Zyban, a nicotine patch, or nicotine gum. According to these findings, snus was nearly three times more effective than nicotine gum, the second most popular means of trying to quit.

The Population Health Effects of THR

Two recent studies looked at the potential population health effects of snus use in populations where the products are not currently available, and they determined that the potential health benefits outweighed the risks.

A 2007 study assessed the potential health gains if snus was available in Australia, where it is currently banned. The researchers calculated the life expectancy among people with various histories of tobacco use, and then estimated the net effects at the population level. The results showed little difference in the life expectancy of smokers who quit all tobacco and smokers who switch to snus. The researchers concluded: "Current smokers who switch to using snus rather than continuing to smoke can realize substantial health gains. Snus could produce a net benefit to health at the population level if it is adopted in sufficient numbers by inveterate smokers."

In another study, researchers looked at the difference between mortality from lung cancer (the sentinel disease of smoking and an indicator of a population's smoking rate) in Sweden, where snus is widely used, and in other European Union countries, where snus is banned. In Sweden, lung cancer mortality was significantly lower among men than in the other EU countries. In fact, if all EU countries had the lung cancer mortality of men in Sweden, there would have been over 53 percent fewer lung cancer deaths in 2002 alone. The researchers found that the number of smoking-attributable deaths would have more than halved if EU smoking rates were similar to those of Swedish men.

A point that further reinforces the relative health benefits of snus is the disparity between smoking rates in Swedish men and women. For most of the last 50 years, lung cancer mortality among Swedish women has been the sixth highest in the EU, reflecting the fact that cigarettes are the dominant tobacco product among women in Sweden. This context is important, because it has been suggested that vigorous anti-smoking campaigns since the 1970s are largely responsible for the decline in the smoking rate among Swedish men. But it is implausible that these campaigns have been effective for Swedish men while having almost no impact on Swedish women. This is firm evidence that snus use — not anti-smoking campaigns — has played the primary role in the low lung cancer mortality among men in Sweden for over half a century.

What Clinical Trials Tell Us About THR

Prior to 2006, only one clinical trial relevant to THR had been conducted; during the last five years, several clinical trials have been completed. In the majority of these trials, researchers evaluated smokers' preferences for a variety of smoking cessation aids, as well as rates of quitting, effects on withdrawal and craving, exhaled levels of carbon monoxide, and levels of tobacco-related metabolites.

In general, snus and dissolvable tobacco satisfied smokers' cravings and helped smokers quit or decrease cigarette consumption, and these products typically were preferred or ranked on par with pharmaceutical nicotine.

American Survey Evidence of Smokeless Tobacco as an Effective Cigarette Substitute

Although the majority of evidence for smokeless tobacco's efficacy comes from Sweden, where consumers are much more aware of smokeless products, there is evidence that it has been effective for small numbers of American smokers.

A 2008 study provided evidence that American men have quit smoking by switching to smokeless tobacco. Using data from the 2000 National Health Interview Survey, the researchers estimated that 359,000 American male smokers had tried to switch to smokeless tobacco during their most recent attempt to quit — and 73 percent of them were former smokers at the time of the survey. These numbers represent the highest proportion of successes among all methods. The researchers found that nicotine gum and the nicotine patch had helped only 34 to 45 percent of male users to quit, while only 28 percent of the men who had tried the nicotine inhaler were successful.

Despite these success rates, few smokers are aware that switching from cigarettes to smokeless tobacco provides almost all the health benefits of complete tobacco abstinence. As one research team reported, risk perception plays an important role in willingness to try snus. In a recent survey, respondents who were aware that smokeless tobacco is less harmful than cigarettes were nearly four times as likely to try snus than those who believed that the products were equally risky. Unfortunately, 88 percent of all respondents in this survey believed that smokeless tobacco was just as dangerous as cigarettes.

Dr. Lois Biener and Dr. Karen Bogen, who reported on this, the Indiana Adult Tobacco Survey, made the following observation:

Both marketing and health education messages should include the information that all tobacco products are harmful and that abstinence from all tobacco products is the most healthful choice. At the same time, simply saying that smokeless tobacco is "not safe" is not a sufficient stance for public health communications. There is a recognized continuum of risk along which various tobacco products can be placed, with low-nitrosamine smokeless tobacco products much lower on the risk continuum than combustible tobacco, although it is not harmless. Devising an effective way to inform the public about the continuum should be an important research priority, as currently consumers are woefully incorrect in their assessments of relative risk of various tobacco products. This state of affairs could result in people deciding not to give up smoking in favor of a product lower on the risk continuum because they assume that all tobacco products are equally harmful.

Smokeless Tobacco Is Not a Gateway to Smoking

A common allegation is that smokeless tobacco is a gateway to smoking cigarettes, especially among youth. However, studies from Sweden and the U.S. discredit this claim.

There is virtually no evidence for a gateway effect in Sweden, where smokeless tobacco use is prevalent. A 2006 review of published studies turned up no evidence, and this was confirmed by a 2008 study of 3,000 adolescents from the Stockholm area: "The majority of tobacco users (70 percent) started by smoking cigarettes," the authors noted. They concluded that the proportion of young people who progressed from smokeless tobacco to cigarettes is small. In addition, the European Commission's Scientific Committee on Emerging and Newly Identified Health Risks concluded that "The Swedish data ... do not support the hypothesis that ... snus is a gateway to future smoking."

In the U.S., opponents of THR believe that it will encourage teenagers to use smokeless tobacco, which will lead to smoking. And, while some studies have reported that teenagers who use smokeless tobacco are more likely to become smokers, a close examination of the evidence suggests only that smokeless tobacco is one of several behaviors associated with smoking — not that it leads to smoking.

The misconception that smokeless tobacco use leads to smoking is based largely on two long-range studies that compared subsequent smoking among adolescent smokeless tobacco users and non-users. While both studies found that young people who use smokeless tobacco were more likely than non-users to be smoking several years later, they neglected to take into account well-known psychological predictors of smoking. When these variables were taken into account, smokeless tobacco use was no longer a statistically significant predictor of developing a smoking habit.

Other recent studies discredited claims of causality. One study found that the association of smokeless tobacco use with smoking is most likely a reflection of experimenting with both substances — not a causal relationship. Another study found that, among white males ages sixteen and older — the group most likely to smoke tobacco — the prevalence of smoking among those who had first tried smokeless tobacco was significantly lower than among those who were cigarette initiators. The researchers concluded that "smokeless tobacco use has played virtually no role in smoking initiation among white men and boys."

Claims of the gateway effect persist, prompting experts at Penn State's tobacco addiction department to note, "Continued evasion of the [harm reduction] issue based on claims that smokeless tobacco can cause smoking seems, to us, to be an unethical violation of the human right to honest, health-relevant information."

Dual Use: Smokeless Tobacco and Cigarettes

The dual use of smokeless tobacco and cigarettes is a major criticism by opponents of THR. Although studies suggest that dual use can benefit an otherwise exclusive smoker, critics complain about adverse consequences and assert that dual users will never quit cigarettes, that they are risking their health just as much as exclusive smokers. However, these assertions are unfounded.

In 2002, a study describing the theoretical adverse consequences of dual use acknowledge that "there are virtually no data that currently exist on the safety of such use or the degree to which such use will foster the perpetuation of smoking or contribute to reduced overall smoking." The researchers concluded, "The issue warrants further study." And indeed, further study has been done. The results are in favor of dual use.

Many smokers gradually begin using smokeless tobacco with the goal of eventually quitting tobacco altogether or, at least, cutting back on their cigarette consumption. Recent studies suggest such a course is not at all misguided.

First of all, the most recent studies suggest that dual users are not increasing their health risks. A 2010 analysis of 17 published research studies concluded that there are no "unique health risks associated with dual use... that are not anticipated or observed from cigarette smoking alone." Furthermore, the authors observed that "some data indicate that the risks of dual use are lower than those of exclusive smoking."

The 2010 analysis also found that dual users are more likely to quit smoking than are exclusive smokers. Longitudinal studies — those that follow their subjects over time — have found that dual users have a different trajectory of tobacco use and cessation than that of exclusive smokers. For instance, a 2002 study in the U.S. found that, after four years, 80 percent of smokers were still smoking, while only 27 percent of those who had been dual users were still smoking. Of the dual users, 44 percent remained dual users and 17 percent used only smokeless tobacco products. A 2003 study in Sweden found similar success rates for dual users; and it found that the rate at which dual users quit smoking entirely increased each year.

These studies made it clear that dual users are much more likely to successfully quit smoking than are exclusive smokers. And, although dual users are less likely than the exclusive smokers to become completely tobacco abstinent, they are much more likely to smoke fewer cigarettes.

Electronic Cigarettes: Another Low-Risk Alternative

E-cigarettes are battery-powered devices that vaporize a mixture of water, propylene glycol, nicotine, and flavorings. They are activated when the user inhales through the mouthpiece of the device, delivering a small dose of nicotine without any of the carcinogens derived from the combustion of tobacco that occurs in cigarettes. To date, all e-cigarettes and mixtures are manufactured in China — although this may change following a recent FDA decision.

In 2011, an appellate court confirmed that e-cigarettes are to be regulated by the FDA as tobacco products, which was a victory on several counts for American smokers and for public health. First, the decision guaranteed that e-cigarettes, which have helped many smokers quit, will remain on the market. Second, it assured that e-cigarettes would be subjected "to general controls, such as registration, product listing, ingredient listing, good manufacturing practice requirements, user fees for certain products, and the adulteration and misbranding provisions, as well as to the premarket review requirements for 'new tobacco products' and 'modified risk tobacco products.'" These requirements will promote the marketing of safe and quality-controlled products. Finally, the decision could allow pharmaceutical companies to reposition more satisfying nicotine medicines as recreational (and low-risk) alternatives to cigarettes.

E-cigarettes: What Studies Tell Us

Clinical and laboratory studies have only begun to accumulate during the last 5 years, and most confirm that e-cigarettes are a safe and effective smoking cessation aid.

While cigarette smoke contains thousands of chemical agents in addition to nicotine, e-cigarettes produce a vapor comprised only of water, propylene glycol, nicotine, and flavorings. Unlike tobacco cigarettes, the ingredients in e-cigarettes do not pose any significant health risks. Nicotine is highly addictive, but it is not the primary cause of any of the diseases related to smoking. Propylene glycol is approved by the FDA for use in a large number of consumer products, and it is not associated with any adverse health effects, although there are currently no studies relating to long term daily exposure.

Despite the relative safety and efficacy of e-cigarettes, tobacco-control activists have aggressively attacked these products. For instance, in 2009, Dr. Jack Henningfield, who is a scientific adviser on tobacco to the World Health Organization, as well as an adviser to GlaxosmithKline on pharmaceutical nicotine, called e-cigarettes "renegade products," for which "we have no scientific information." He then stated that e-cigarettes "are not benign" — although there was no explanation in his article as to how he came to that conclusion in the absence of any scientific information.

While it is true that there is a paucity of scientific studies on this new area, and that the discussion has become highly polarized, several reports have provided important information.

Clinical Studies

Studies that have addressed how the nicotine in e-cigarettes is absorbed suggest that it is largely taken up by the mucous membranes of the mouth and throat, and that this is typically achieved by shallow puffing instead of deep inhalation. Significantly, a study has also found that e-cigarette use results in much less frequent mouth and throat irritation, compared to the pharmaceutical nicotine inhaler. And, just as important, another study has found that, although the e-cigarette produces only modest elevations in peak blood levels of nicotine — much lower than that produced by cigarettes — users experienced both reduced cravings and withdrawal symptoms. Researchers believe that this reduction is due not solely to nicotine delivery but to the e-cigarettes' successful mimicry of cigarette handling rituals and cues — a feature that is not part of pharmaceutical nicotine products.

Laboratory Studies

Although it is clear that the vapor produced by e-cigarettes is simply not comparable to the thousands of toxic agents formed when tobacco is burned, those opposed to e-cigarettes often target their safety. While laboratory studies have detected trace levels of some contaminants in these devices, this appears to be a small problem that could be solved with improvements to quality and manufacturing that will come with FDA regulation.

Unfortunately, media attention typically gravitates to a study released by the FDA in 2009, which stated that its laboratory tests of e-cigarettes "indicated that these products contained detectable levels of known carcinogens." However, this study was flawed and not an accurate indicator of the safety of e-cigarettes.

The FDA examined only a small sample size of e-cigarette cartridges and did not conduct the testing in a systematic and scientific manner: All in all, the agency ended up testing only ten e-cigarette cartridges from only two different manufacturers. It would be difficult to conclude anything from such a tiny sampling of products.

To further complicate matters, the FDA tested e-cigarettes for carcinogens called tobacco-specific N-nitrosamines (TSNAs), but they did not report the levels they detected. Instead, the agency merely reported that TSNAs were either "detected" or "not detected," which is uninformative. Many tobacco products, including smoking cessation aids such as nicotine medications, have TSNA levels in the single-digit parts per million range — a level at which there is no scientific evidence that they are harmful. What's more, the FDA tested for TSNAs using a method that detects TSNAs at about one million times lower concentrations than are conceivably related to human health.

In sum, the FDA tested e-cigarettes for TSNAs using a questionable sampling regimen and using methods that were so sensitive that the results are highly unlikely to have any possible significance to users. Thus far, laboratory analysis suggests that e-cigarettes do not contain harmful levels of carcinogens. A much more comprehensive study, employing much more precise methods than those used by the FDA, would go a long way toward clarifying the relative safety of e-cigarettes.

The Growing Global Discussion About THR

There is growing global interest in THR. More and more studies are confirming the safety of smokeless products, leading to a greater awareness of the option, as well as more public discussion.

In addition to ACSH's conclusion that "there is a strong scientific and medical foundation for THR, which shows great potential as a public health strategy," many other experts are voicing their convictions in favor of harm reduction.

In 2007, the Royal College of Physicians, one of the oldest and most prestigious medical societies in the world, published a landmark report, concluding that,"if nicotine could be provided in a form that is acceptable as a cigarette substitute, millions of lives could be saved."

Furthermore, in a variety of scientific journals, researchers are underscoring the relatively low risks of smokeless products and are calling for them to be incorporated into official strategies for helping smokers to quit.

Two Australian researchers, Coral Gartner and Wayne Hall, wrote in a 2007 *Public Library of Science Medicine* article that the health risk of smokeless tobacco are "comparable to those of regular alcohol use rather than cigarette smoking." They called for a public health policy that would promote the use of smokeless products for inveterate smokers.

And, perhaps it's best to close with the following summary of the global health implications of THR:

> The relative safety of ST and other smokefree systems for delivering nicotine demolishes the claim that abstinence-only approaches to tobacco are rational public health campaigns... Applying harm reduction principles to public health policies on tobacco/nicotine is more than simply a rational and humane policy. It is more than a pragmatic response to a market that is, anyway, already in the process of undergoing significant changes. It has the potential to lead to one of the greatest public health breakthroughs in human history by fundamentally changing the forecast of a billion cigarette-caused deaths this century.

— David T. Sweanor, former Senior Legal Counsel to the Smoking and Health Action Foundation et al., *International Journal of Drug Policy* (2007)

BOARD OF SCIENTIFIC AND POLICY ADVISORS *(continued)*

BOARD OF SCIENTIFIC AND POLICY ADVISORS *(continued)*

BOARD OF SCIENTIFIC AND POLICY ADVISORS *(continued)*

BOARD OF SCIENTIFIC AND POLICY ADVISORS *(continued)*

BOARD OF SCIENTIFIC AND POLICY ADVISORS *(continued)*

Made in the USA
Monee, IL
04 November 2026

67386132R00020